I0478536

Natural Science Series Volume 1

Molecular Mechanics of the Ideal Gas

Hiroyuki Aizawa

Aizawa Science Museum
Yutaka-cho 1-10-13
Kasukabe, Saitama 3440066, JAPAN
Tel: +81-48-754-9880
e-mail: aizawa@rr.iij4u.or.jp
http://www.aizawa.com/newpage6.html

Foreword

When a wind blows, we feel cool but not hot. The electric fan accelerates air molecules but the room temperature is not increased by the wind. The wind is not so hot as expected. This experience encouraged me to develop a new theory of thermodynamics.

The vision of subjects in the world appears so clear to our eyes, suggesting that the wave of light should travel straight in the atmosphere with air molecules. Since the vision is sometimes perturbed by a heat haze but not by a wind, thermodynamic nature of the air should affect the speed of light.

These considerations incited me to postulate that a gas molecule is a rotating elastic sphere contacting with each other. The temperature of air should be rotation energy of gas molecules, and a wave of light should travel all along the gas molecules that act as media of light.

In this tiny book, I would postulate two axioms for development of novel thermodynamics. I hope that the postulation may help not only to explain the well-known law of the ideal gas as shown here but also to develop physical optics in the future.

Contents

Introduction

Kinetic theory of gases has suggested that a tiny rigid spherical body travels freely in space with occasional collision against each other in a perfectly elastic manner. The theory explains wave mechanics of sound well but not that of light, suggesting the necessity of its improvement for optics. Here, we postulate that an ideal gas molecule is an elastic body, which becomes a shell-like sphere by centrifugal force derived from its rotation. We also postulate that the rotating sphere contacts to each other. Molecular mechanics of the ideal gas develops mathematical expression for the law of the ideal gas. The theory indicates that the rotation energy of a gas molecule is in proportion to a product of pressure and volume, and thus creates space. This theory develops novel mechanics of the ideal gas molecules.

Axiom

Axiom One
A molecule is an elastic body.

Axiom Two
Molecules contact with each other.

Definition

Definition One
Modulus of elasticity is a negatively proportional coefficient of the differential stress to the differential deformation of an elastic body.

Modulus of elasticity of a linear body with a unit of cross-sectional area is expressed as follows:
$$k_l = - df_k / dx$$
where k_l, df_k, and dx indicate modulus of elasticity per a unit of cross-sectional area ($kg \cdot m^{-2} \cdot s^{-2}$), differential stress ($kg \cdot m^{-1} \cdot s^{-2}$), and differential deformation (m), respectively.

Thus, modulus of volume elasticity, which corresponds to modulus of elasticity of a body with a unit length per a unit of cross-sectional area, is expressed as follows:
$$k_v = k_l \cdot l$$
where k_v and l indicate modulus of volume elasticity ($kg \cdot m^{-1} \cdot s^{-2}$) and length (m) of a body, respectively.

Definition Two
Modulus of proper elasticity is the modulus of elasticity of a body with a unit of mass in a unit of volume.

From Definition One, modulus of proper elasticity is expressed as follows:
$$\kappa_v = k_v / \rho_v$$

where κ_v and ρ_v indicate modulus of proper elasticity ($m^2 \cdot s^{-2}$) and volume density (kg/m^3), respectively. Modulus of proper elasticity is a constant value characteristic to each substance. Thus, modulus of proper elasticity per area and modulus of proper elasticity per length are expressed as follows:

$$\kappa_s = \kappa / l \; (l = 1 \text{ m})$$
$$\kappa_l = \kappa / S \; (S = 1 \text{ m}^2)$$
$$k_v = \kappa_v \cdot \rho_v = \kappa_s \cdot \rho_s = \kappa_l \cdot \rho_l$$

where κ_s, κ_l, ρ_s, ρ_l, l, and S indicate modulus of proper elasticity per area ($m \cdot s^{-2}$), modulus of proper elasticity per length (s^{-2}), density per area (kg/m^2), density per length (kg/m), a unit of length (m), and a unit of area (m^2) of the body, respectively. The numbers of κ_v, κ_s, and κ_l are the same while the units of them are unique to each modulus of elasticity. Accordingly, in this book we use characters "κ" and "ρ" as modulus of proper elasticity and density, respectively, for convenience as follows:

$$k_v = \kappa \cdot \rho$$

Definition Three
Molecular weight or relative molecular mass is a ratio of mass of a molecule to a unified atomic mass unit.

Molecular weight is dimensionless quantity. A unified atomic mass unit corresponds to 1.66 x 10^{-27} kg. Molecular mass is a product of molecular weight and a unified atomic mass unit. Avogadro's constant ($N_A = 6.02$ x 10^{23}) is the number of molecules chemical compounds, whose mass in gram corresponds to its molecular weight.

Definition Four
An ideal gas molecule is an elastic spherical shell-like body, which is formed by centrifugal force of rotation in excess to the maximal stress of the elastic body.

Definition Five
Heat of an ideal gas molecule is torque or momentum of force applying to each other.

Definition Six
Quantity of heat of an ideal gas molecule is its rotation energy.

Rotation energy is the integration of torque along angle. The rotation energy of a spherical shell is expressed as follows:
$$\varepsilon_R = (1/3) \, m \cdot r^2 \cdot \omega^2$$
where ε_R, m, r, and ω indicate rotation energy ($kg \cdot m^2 \cdot s^2$), mass (kg), radius (m), and angular speed ($rad \cdot s^{-1}$), respectively.

Definition Seven
Heat capacity of an ideal gas molecule is the workload necessary to increase its quantity of heat by 1/273.15 of

that at the freezing point of water.

Definition Eight
An adiabatic process is a thermodynamic process under a constant angular momentum of a gas molecule.

Angular momentum of a spherical shell is expressed as follows:

$$L = (2/3)\, m \cdot r^2 \cdot \omega$$

where L indicates angular momentum ($kg \cdot m^2 \cdot s^{-1}$) of a gas molecule.

Theorem

Theorem One

When a linear elastic body of a mass "m" (kg) with a unit of cross-sectional area (m²) and an original length "r_0" (m) is deformed to length "r" (m) with a unit cross-sectional area by external force, stress "$f_k(r)$" (kg·m^{-1}·s^{-2}) at an end of the body is expressed as follows:

$$f_k = \kappa \cdot m \cdot (1/r - 1/r_0)$$

where κ indicate modulus of linear proper elasticity of an elastic body (m²·s^{-2}).

Figure 1   Relationship of length (r) and stress (f_k)
of a linear elastic body.

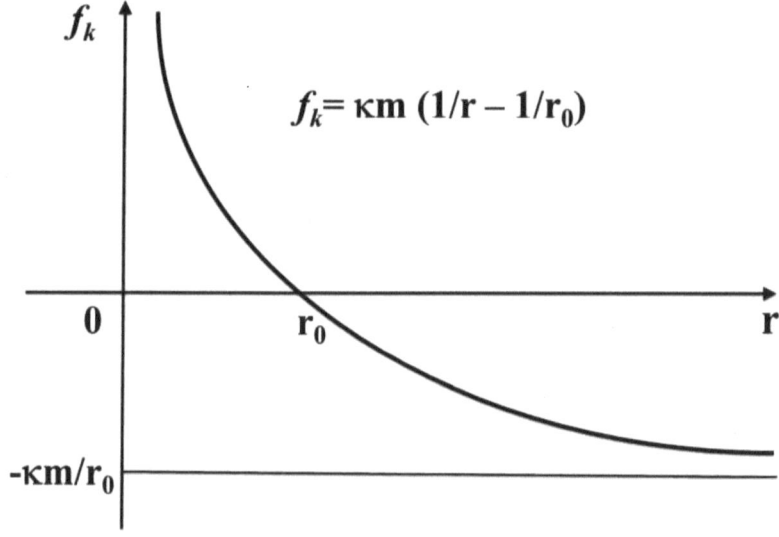

Proof One

From Definition One, differential stress "dfk" of a linear elastic body at length of "r" is expressed as follows:

$$df_k = -k_l \, dr$$

where k_l and dr indicate modulus of elasticity and differential length of the body, respectively. Accordingly, the differential stress is expressed as follows:

$$df_k = -(k_v/r)dr$$

where k_v indicates modulus of a volume elasticity of a linear body. From Definition Three, the differential stress is also expressed as follows:

$$df_k = -(\kappa \cdot m/r^2) \, dr$$

where κ indicates modulus of proper elasticity of the body. Stress of the linear elastic body at length of r is calculated by definite integration of differential stress from r_0 to r, and thus expressed as follows:

$$f_k(r) = \kappa \cdot m \cdot (1/r - 1/r_0)$$

where $f_k(r)$ indicate stress of the linear elastic body at length of r (m).
[End of Proof]

Thus, contractile stress of the body takes its maximal value when r becomes infinity as follows:

$$-f_k^{max} = \kappa \cdot m / r_0$$

where $-f_k^{max}$ indicates the maximal value of contractile stress.

On the other hand, repulsive force increases larger as length of the body is compressed to become shorter than that of original one, and the repulsive force becomes infinity when the length of body becomes close to zero.

According to Definition Two, modulus of elasticity is in

reverse proportion to square of length of the body and thus expressed as follows:

$$k_l = \kappa \cdot m / r^2$$

When length of the body becomes infinity, modulus of elasticity becomes zero and then the length could be easily modified by a tiny difference of external force applied to the body. For example, we need to apply relatively large pressure to start blowing up a rubber balloon while we need to add only a small pressure to enlarge a big balloon. This is because modulus of elasticity becomes smaller as the body becomes longer.

Theorem Two

When linear density at position x "$\rho(x)$" of a linear elastic body with a unit of cross-sectional area is not a function of time but a function of position "x" from an end, around which the body is rotating at angular speed of ω, the density "$\rho(x)$" is expressed as follows:

$$\rho(x) = A \cdot \exp\{(\omega^2 / 2\kappa) \cdot x^2\}$$

where κ, m, r, and r_0 indicate modulus of proper elasticity (s^{-2}), mass (kg/m^2), rotating length (m), and original length (m) of the body, respectively, and A is expressed with a definite integral "$\int$" from zero to r as follows:

$$A = m / \{\int \exp(\omega^2 \cdot x^2 / 2\kappa) dx\}$$

Proof Two

Centrifugal force "$f_c(x)$" and stress "$f_k(x)$" applying on a body within a differential distance dx at position x of a linear elastic body are expressed as follows:

$$f_c(x) = x \cdot dx \cdot \rho(x) \cdot \omega^2$$
$$f_k(x) = -\kappa \cdot \{\rho(x+dx) - \rho_0\} + \kappa \cdot \{\rho(x) - \rho_0\}$$
$$= -\kappa \cdot \{\rho(x+dx) - \rho(x)\}$$

where dx and ρ_0 indicate differential of position x (m) and density (kg/m^3) of the body at original length of r_0. If the rotating linear elastic body keeps its length at a constant value, its centrifugal force is balanced by stress at any position x of the body as follows:

$$f_c(x) + f_k(x) = 0$$

Accordingly, density $\rho(x)$ is expressed as follows:

$$\omega^2 \cdot x \cdot \rho(x) = \kappa \cdot \{\rho(x+dx) - \rho(x)\} / dx = \kappa \cdot d\rho(x)/dx$$
$$\therefore \rho(x) = A \cdot \exp\{(\omega^2 /2\kappa) \cdot x^2\}$$

where A is a constant value. Since integral of density from 0 to r is "m" (kg/m^2), A is expressed as follows:

$$A \cdot \{\int \exp(\omega^2 \cdot x^2/2\kappa)dx\} = m$$
$$\therefore A = m / \{\int \exp(\omega^2 \cdot x^2/2\kappa)dx\}$$

where $\int$ indicates integral from 0 to r.
[End of Proof]

Theorem Three
When a linear elastic body with modulus of elasticity "κ", original length "r_0", and mass (kg) per a unit of cross-sectional area (m^2) is rotating around an end at an angular speed of ω to deform into a linear body with constant length "r", density at position x "$\rho(x)$" of the body is expressed as

follows:

$$\rho(x) = (m/r_0) \cdot \exp\{(\omega^2/2\kappa) \cdot (x^2 - r^2)\}$$

Proof Three

Centrifugal force "$f_c(x)$" and stress "$f_k(x)$" applying on a body within a differential distance dx at the distal end of a rotating body are expressed as follows:

$$f_c(x) = r \cdot dx \cdot \rho(x) \cdot \omega^2$$
$$f_k(x) = \kappa \cdot \{\rho(r-dx) - \rho_0\}$$

where dx and ρ_0 indicate differential of position x (m) and density (kg/m³) of the body at original length of r_0. When the centrifugal force is balanced by stress to maintain the length of the body at length of "r", summation of forces at r becomes 0 as follows:

$$f_c(r) + f_k(r) = r \cdot dx \cdot \rho(r) \cdot \omega^2 + \kappa \cdot \{\rho(r-dx) - \rho_0\} = 0$$

When dx becomes close to 0, the equation above becomes as follows:

$$\kappa \cdot \{\rho(r) - \rho_0\} = 0$$
$$\therefore \rho(r) = \rho_0 \ (= m/r_0)$$

Accordingly, from Theorem Two,

$$A = (m/r_0) \cdot \exp\{(\omega^2/2\kappa) \cdot (-r^2)\}$$

Thus, density distribution of the linear elastic body along its length is expressed as follows:

$$\rho(x) = (m/r_0) \cdot \exp\{(\omega^2/2\kappa) \cdot (x^2 - r^2)\}$$

[End of Proof]

Theorem Four

In Theorem Two, if an external pressure is applied at the distal end of the rotating elastic body to deform it with constant length of "r", the density of the body is expressed as follow:

$$\rho(x) = (m/r_0 - p/\kappa)\cdot\exp\{(\omega^2/2\kappa)\cdot(x^2-r^2)\}$$

where "p" is an external pressure (kg·m^{-1}·s).

Figure 2
Density distribution of a rotating elastic body under pressure

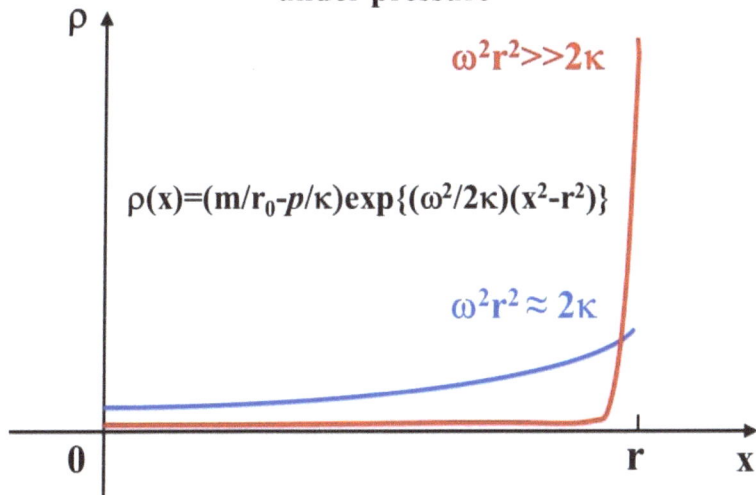

Proof Four

Centrifugal force "$f_c(x)$", stress "$f_k(x)$", and an external force "$f_p(x)$" as pressure "p" applying on a rotating body within a differential distance dx are expressed as follows:

$$f_c(x) = x\cdot dx\cdot\rho(x)\cdot\omega^2$$
$$f_k(x) = -\kappa\cdot\{\rho(r+dx) - \rho_0\} + \kappa\cdot\{\rho(r) - \rho_0\}$$
$$= -\kappa\cdot\{\rho(x+dx) - \rho(x)\}$$

$$f_p(x) = -p + p = 0$$

where dx, ρ_0, and p indicate differential of position x (m), linear density per unit of area (kg/m³) of the body at original length of r_0, and pressure (kg/m⁻¹/s⁻²). If the rotating linear body keeps its length at a constant value, its centrifugal force is balanced by stress at any position x of the body as follows:

$$f_c(x) + f_k(x) + f_p(x) = 0$$

As Theory Two, density $\rho(x)$ is expressed as follows:

$$\rho(x) = A \cdot \exp\{(\omega^2/2\kappa) \cdot x^2\}$$

A is expressed as follows:

$$A = m / \{\int \exp(\omega^2 \cdot x^2/2\kappa)dx\}$$

where $\int$ indicates integral from 0 to r.

When external force is balanced by stress and centrifugal force to maintain the length of the body at constant length "r", summation of forces at r should become 0 as follows:

$$f_c(r) + f_k(r) + f_p(r) = r \cdot dx \cdot \rho(x) \cdot \omega^2 + \kappa \cdot \{\rho(r\text{-}dx) - \rho_0\} + p = 0$$

Here, if dx becomes close to 0, the equation above becomes as follows:

$$\kappa \cdot \{\rho(r) - \rho_0\} = -p$$
$$\therefore \rho(r) = \rho_0 - p/\kappa = m/r_0 - p/\kappa$$

As mentioned above, density at r is also expressed as follows:

$$\rho(r) = A \cdot \exp\{(\omega^2/2\kappa) \cdot r^2\}$$

Thus, A is expressed as follows:

$$A = (m/r_0 - p/\kappa) \cdot \exp\{(\omega^2/2\kappa) \cdot (-r^2)\}$$

16

Consequently, density distribution of the linear elastic body is expressed as follows:

$$\rho(x) = (m/r_0 - p/\kappa)\cdot\exp\{(\omega^2/2\kappa)\cdot(x^2-r^2)\}$$

[End of Proof]

Theorem Five

In Theorem Two to Four, a square of a product of angular speed and length of the rotating linear body "$\omega^2\cdot r^2$" becomes much larger than double of proper elasticity of the body "2κ", the density at the rotation center "$\rho(0)$" becomes close to zero.

Proof Five

From Theorem Two to Four, the density at the rotation center is expressed as follows:

$$\rho(0) = A = m \ / \int\exp(x^2\cdot\omega^2/2\kappa)dx$$

Here, differentiation of the integral by r becomes infinity as "$r^2\cdot\omega^2/2\kappa$" increases to infinity as follows:

$$d\{\int\exp(x^2\cdot\omega^2/2\kappa)dx\} \ / \ dr = \exp(r^2\cdot\omega^2/2\kappa) \rightarrow \infty$$
$$(\text{when } r^2\cdot\omega^2/2\kappa \rightarrow \infty)$$

Thus, as shown in Figure 2,

$$\int\exp(x^2\cdot\omega^2/2\kappa)dx \rightarrow \infty \ (\text{when } r^2\cdot\omega^2/2\kappa \rightarrow \infty)$$
$$\therefore \rho(0) = A \rightarrow 0 \ (\text{when } r^2\cdot\omega^2/2\kappa \rightarrow \infty)$$

Thus, the density at the rotation center becomes close to zero when a square of a product of angular speed and length of the rotating linear body "$\omega^2\cdot r^2$" becomes much larger than double of proper elasticity of the body "2κ".

[End of Proof]

Thus, if an elastic body rotates around at an angular speed much higher than the square root of $2\kappa/r^2$, its rotation central part becomes almost empty.

Theorem Six

In Theorem Four, when a product of square of angular speed and length of the rotating linear body "$\omega^2 \cdot r$" becomes much larger than proper elasticity κ of a body, almost all the mass of the body accumulates at its external end part, where the density is expressed as follows:

$$\rho(r) = m/r_0 - p/\kappa$$

Proof Six

From Theory Four, differentiation of density $\rho(x)$ is expressed as follows:

$$d\rho/dx = (m/r_0 - p/\kappa) \cdot (\omega^2/\kappa) \cdot x \cdot \exp\{(\omega^2/2\kappa) \cdot (x^2 - r^2)\}$$

Thus, at position r, differentiation of density is expressed as follows:

$$d\rho(r)/dx = (m/r_0 - p/\kappa) \cdot (r \cdot \omega^2/\kappa)$$

Consequently, when angular speed ω increases till infinity, the differentiation of density at position r becomes infinity, suggesting that almost all the mass accumulate within a differential distance dx from the distal end of the rotating body.

From Theory Four, the density at the end is expressed as

18

follows:
$$\rho(r) = m/r_0 - p/\kappa$$
[End of Proof]

Because pressure p is a negative value to decrease the length of a body, density at a distal end of a rotating body becomes larger than ρ_0 (= m/r_0) under pressure. If the pressure becomes negative infinity at a distal end, the density at the distal end increases to become infinity.

Theorem Seven
In Theorem Five, if external force "f_P" or pressure "p" balances with centrifugal force "f_c" and stress "f_k" of the elastic body to maintain a length of a rotating linear body at "r", the expression of p is as follows:
$$p = -m \cdot r \cdot \omega^2$$
where "m" is mass of the linear body per a unit of cross-sectional area (kg/m^2).

Proof Seven
From Theory Four, center of mass of a rotating linear body with length of r around its end "r_c" is expressed as follows:
$$r_c = \{\int x \cdot \rho(x)dx\} / m$$
$$\Leftrightarrow r_c = \{(\kappa/r_0/\omega^2 - p/m/\omega^2)\} \cdot \{1 - \exp(-\omega^2 \cdot r^2/2\kappa)\}$$
According to Proof Six, center of mass locates closer to a distal end of the linear body as an angular speed ω becomes higher. Consequently, when an angular speed increases

higher till infinity, center of mass is expressed as follows:

$$r = - p/m/\omega^2 \text{ (when } r^2 \cdot \omega^2/2\kappa \to \infty)$$

$$\therefore p = -m \cdot r \cdot \omega^2$$

[End of Proof]

Since almost all the mass of a body rotating around an end locates at the distal end, the centrifugal force of the body f_c is expressed as follows:

$$f_c = m \cdot r \cdot \omega^2 \text{ (when } r^2 \cdot \omega^2/2\kappa \to \infty)$$

where m, r, and ω indicate mass per a square unit (kg·m^{-2}), length (m), and an angular speed (rad·s^{-1}), respectively. When the body keeps a constant length during rotation, the centrifugal force is balanced with an external force or pressure "p", which is thus expressed as follows:

$$p = - f_c = - m \cdot r \cdot \omega^2 \text{ (when } r^2 \cdot \omega^2/2\kappa \to \infty)$$

When $r^2 \cdot \omega^2$ is much larger than 2κ, we are going to consider an elastic linear body rotating around an end as a mass point with a mass per unit square "m" (kg·m^{-2}) rotating around a center with radius "r" (m) at an angular speed "ω" (rad·s^{-1}).

Theorem Eight

In Theorem Six, rotation energy "ε_R" of the linear body is expressed as follows:

$$\varepsilon_R = (1/2) \, m \cdot r^2 \cdot \omega^2$$

Proof Eight

From Theorem Seven, almost all the mass "m" of the rotating body locates at the distal end at position "r". Thus, the moment of inertia of the body I is expressed as follows:

$$I = mr^2$$

Accordingly, the rotational energy ε_R is expressed as follows:

$$\varepsilon_R = (1/2)\, I \cdot \omega^2 = (1/2)\, m \cdot r^2 \cdot \omega^2$$

[End of Proof]

Theorem Nine

An elastic sphere rotating around its center of mass with randomly moving rotational vector behaves as an elastic spherical shell called an ideal gas molecule, when square of product of time-averaged angular speed and radius of the spherical shell is much larger than double of modulus of proper elasticity of the body.

Proof Nine

From Theorem Five and Six, when a square of a product of angular speed and length of the rotating body "$\omega^2 \cdot r^2$" becomes much larger than double of proper elasticity of the body "2κ", the central inner mass of the elastic body becomes almost empty and almost all the mass locates at the distal end of the rotating body. Thus, a rotating elastic sphere becomes a spherical shell-like body, which is called an ideal gas molecule from Definition Four.

[End of Proof]

Now we are shifting our mechanical model of elastic body from linear one to sphere. When an elastic sphere rotates around its center of mass, centrifugal force directs radially from and perpendicular to the rotational axis. Accordingly, the sphere tends to become a disc-like shape. If the rotational axis randomly rotates around the center of mass by interaction with neighboring molecules, the time-averaged rotational vector becomes zero while time-averaged angular speed becomes ω. The rotatiing body becomes time-averaged spherical shell-like shape with a mass "m" and an averaged radius "r" while at each time point the body took an undefined spindle-like shape. This is the mechanical model of the ideal gas molecule. We will study on the molecular mechanics of the ideal gas in detail below.

Theorem Ten
Under pressure "P" of atmosphere, when an elastic sphere with mass "m" is rotating around its center of mass with a time-averaged angular speed vector and its averaged absolute value at 0 and ω, respectively, with which the elastic sphere almost becomes a spherical shell of radius "r", the pressure "P" is expressed as follows:

$$P = m \cdot \omega^2 / 6\pi r$$

Figure 3 Spherical shell model of an ideal gas molecule

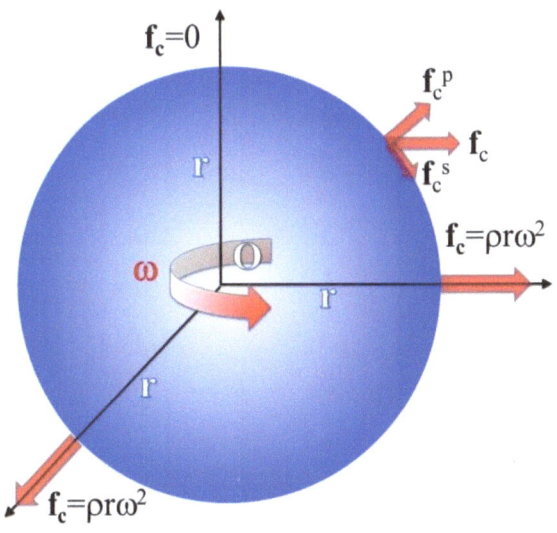

Proof Ten

The centrifugal force of each point on surface of the rotating spherical shell is expressed as follows:

$$f_c = \rho_s \cdot r \cdot \omega^2$$

where f_c, ρ_s, r, and ω is the centrifugal force per a unit of area ($kg \cdot m^{-1} \cdot s^2$), surface density ($kg \cdot m^{-2}$), radius (m), and time-averaged angular speed ($rad \cdot s^{-1}$), respectively. The surface density is expressed by mass and radius of the body as follows:

$$\rho_s = m / 4\pi r^2$$

where m indicates mass (kg) of the body. Thus, f_c is expressed as follows:

$$f_c = m \cdot \omega^2 / 4\pi r$$

Consequently, a space-averaged radial component of the

centrifugal force of the ideal gas molecule is expressed as follows:

$$f_c^p = \iint \rho_s \cdot r \cdot \omega^2 dS / S$$
$$\Leftrightarrow f_c^p = \iint \rho_s \cdot r^2 \cdot \sin^2\theta \cdot r \cdot \sin\theta \cdot \omega^2 \cdot d\theta \cdot d\varphi / 4\pi r^2$$
$$\Leftrightarrow f_c^p = (2/3)\, \rho_s \cdot r \cdot \omega^2$$
$$\therefore\ f_c^p = m \cdot \omega^2 / 6\pi r$$

where f_c^p, ρ_s, r, ω, S, and m indicate an averaged radial component of centrifugal force (kg·m·s^{-2}), surface density (kg·m^{-2}), radius (m), averaged angular speed (rad·s^{-1}), surface area (m^2), and molecular mass (kg) of the body, respectively.

From Theorem Six, almost all the mass of an elastic body accumulates at the position r from the center of rotation under a pressure "P" when f_c^p becomes equal to "P". Thus, P is expressed as follows:

$$P = m \cdot \omega^2 / 6\pi r$$

[End of Proof]

Thus, P is also expressed as follows:

$$P = -f_p = -p$$

where f_p and p indicate external force and pressure applied onto the surface of the ideal gas molecule.

As shown in Figure 2, centrifugal force of rotational spherical shell could be divided into two components, that is, radial one and surface one. The former gives pressure of atmosphere to neighbor spherical shells while the latter gives torque as heat. Only the pressure in the two could work for internal combustion engine, and this is the reason that engines heats up by the torque when the pressure works to push pistons. Accordingly, the maximum efficiency of the internal combustion engine is 2/3 or 66.7% of the total energy in gas molecules, while torque works for rotational motion of the gas molecules as heat and could be translated into other types of energy such as electricity, light, sound, elastic potential by some proper conversion methods.

Theorem Eleven

In Theorem Ten, rotation energy "ε_R" of a spherical shell is expressed by pressure "P" and space and time-averaged volume "υ" as follows:

$$P \cdot \upsilon = (2/3)\ \varepsilon_R$$

Proof Eleven

The moment of inertia "I" and rotational energy "ε_R" of an ideal gas molecule with mass "m", radius "r", and volume "υ" are expressed as follows:

$$I = (2/3)m \cdot r^2$$
$$\varepsilon_R = (1/2)I \cdot \omega^2$$

And, the volume is expressed by r as follows:

$$\upsilon = (4/3)\pi r^3$$

From Theory Ten, pressure "P" is expressed as follows:

$$P = m \cdot \omega^2 / 6\pi r$$

Thus, product of pressure "P" and volume "υ" are expressed as follows:

$$P \cdot \upsilon = (4/3) \, \pi r^3 m \cdot \omega^2 / \, 6\pi r$$
$$\Leftrightarrow P \cdot \upsilon = (2/9) \, m \cdot r^2 \cdot \omega^2$$
$$\Leftrightarrow P \cdot \upsilon = (2/3) \, (1/2) \, (2/3) \, m \cdot r^2 \cdot \omega^2$$
$$\Leftrightarrow P \cdot \upsilon = (2/3) \, (1/2) \, I \cdot \omega^2$$
$$\therefore \; P \cdot \upsilon = (2/3) \cdot \varepsilon_R$$

[End of Proof]

This is exactly the Boyle's law in case of a single molecule. If internal energy of a gas is constant, its pressure and volume is in reverse proportion to each other. It should be noted that molecular weight of a gas molecule is included in the moment of inertia, which could be modified by alteration of radius of a molecule by rotational movement. Accordingly, product of pressure and volume of an ideal gas molecule definitely depends on its rotational energy but not solely on its molecular weight. Space is created by rotation energy of gas molecules.

Theorem Twelve

Rotation energy of 1 mol of the ideal gas "E_R" is expressed by pressure "P" and volume "V" of the gas as follows:
$$P \cdot V = (2/3) \, E_R$$

Figure 4 Molecular model of an ideal gas

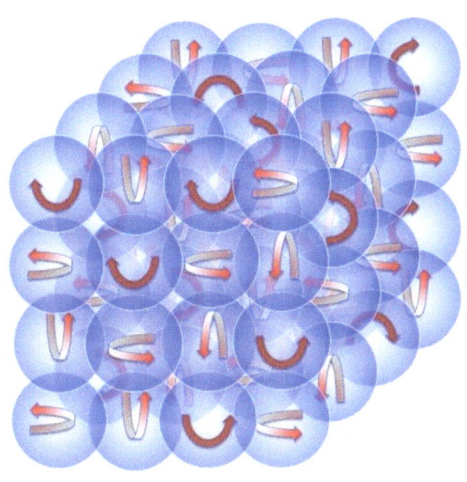

Proof Twelve

When both sides of the equation of Theorem Eleven are multiplied by Avogadro Number "N_A", we have an equation as follows:

$$P \cdot \upsilon \cdot N_A = (2/3)\varepsilon_R \cdot N_A$$

Total volume of 1 mol gas "V" is summation of volume of Avogadro's Number of gas molecules. Since volume "υ" of the ideal gas molecule is time and space-averaged one, total volume is expressed as follows:

$$V = \upsilon \cdot N_A$$

Total energy of 1 mol gas "E_R" is summation of energy of Avogadro's Number of gas molecules. Since energy "ε_R" of the ideal gas molecule is time and space-averaged one, total energy is expressed as follows:

$$E_R = \varepsilon_R \cdot N_A$$

Thus, we have an equation as follows:

$$P \cdot V = (2/3)\, E_R$$

[End of Proof]

Here in this tiny book, we would not develop the mechanics of energy transfer among gas molecules in detail, and thus we do not know the statistical physics of gas molecules.   Still, it is reasonable to speculate that every molecule in a gas has a rotational energy with a standard distribution in space at a single point of time.   It also means that internal energy is distributed evenly among all the gas molecules in a time-averaged manner.   Again, molecular weight of each gas molecule does not affect the energy distribution even in a gas mixed with different kinds of molecules.

Theorem Thirteen

Rotational energy "E_R" of 1mol of the ideal gas is in proportion to temperature "T" with proportional coefficient of C_v as follows:

$$E_R = C_v \cdot T$$

where C_v is the heat capacity of 1mol of the ideal gas under constant volume.

Proof Thirteen

According to Definition Seven, a unit of temperature "T"

corresponds to the 1/273.15 of heat quantity of an ideal gas at freezing point of water.   From Definition Six, rotational energy "E_R" is a heat quantity, which is a product of heat capacity under constant volume "C_v" and temperature "T". Consequently, rotational energy "E_R" is expressed as follows:

$$E_R = C_v \cdot T$$

[End of Proof]

At the freezing point of water, 1mol of an ideal gas under 1atm or $101,325 \ kg \cdot m^{-1} \cdot s^{-2}$ has a volume of 22.4 L or $0.0224 \ m^3$.   From the Theorem Nine, 1mol of an ideal gas has a rotation energy of $3/2 \times 2270 \ kg \cdot m^2 \cdot s^{-2} \cdot mol^{-1}$, which should correspond to $273.15 \ C_v \ kg \cdot m^2 \cdot s^{-2} \cdot mol^{-1}$.   Thus, C_v of an ideal gas is calculated to be $12.5 \ kg \cdot m^2 \cdot s^{-2} \cdot mol^{-1} \cdot K^{-1}$.

Theorem Fourteen
The product of pressure "P" and volume "V" of 1mol of the ideal gas is in proportion to temperature "T" with proportional coefficient of R as follows:

$$P \cdot V = (2/3) \ C_v \cdot T$$

Proof Fourteen
From Theorem Twelve and Thirteen, product of pressure and volume of 1mol of an ideal gas is expressed as follows;

$$P \cdot V = (2/3) \ E_R = (2/3) \ C_v \cdot T$$

[End of Proof]

Thus, product of pressure and volume of the ideal gas is in proportion to temperature with proportional coefficient of R, which is expressed as follows:

$$R = (2/3) \, C_v$$

where R indicates gas constant ($8.3 \text{ kg} \cdot \text{m}^2 \cdot \text{s}^{-2} \cdot \text{K}^{-1} \cdot \text{mol}^{-1}$). This is "the law of the ideal gas" which has been established in the nineteenth century from experiments performed in the seventeenth and eighteenth centuries in the Europe.

Theorem Sixteen

Temperature of an ideal gas molecule is expressed as follows:

$$T = (2/9k_B) \, m \cdot r^2 \cdot \omega^2$$

where k_B is the Boltzmann's constant.

Proof Sixteen

From Theorem Fourteen, rotation energy of an ideal gas molecule is expressed as follows:

$$\varepsilon_R = (3/2) \, k_B \cdot T$$
$$k_B = (2/3)C_v/N_A$$
$$\varepsilon_R = E_R/N_A$$

where N_A is the Avogadro's Number or 6.02×10^{23}. From Definition Seven, temperature is expressed as follows:

$$T = (2/9k_B) \, m \cdot r^2 \cdot \omega^2$$

[End of Proof]

Accordingly, temperature is the rotational energy, which could be altered by torque and pressure. The former alters angular momentum while the latter keeps angular momentum at constant in an adiabatic process as shown below.

Theorem Seventeen
In adiabatic process, 1mol of the ideal gas obey Poisson's low as follows:
$$P \cdot V^{5/3} = \text{Constant}$$
$$T \cdot V^{2/3} = \text{Constant}$$
where P, V, and T indicate its pressure ($kg \cdot m^{-1} \cdot s^{-2}$), volume ($m^3$), and temperature (K), respectively.

Proof Seventeen
From Definition Eight, angular momentum of the ideal gas molecule is constant in an adiabatic process. Thus, angular momentum "L", pressure "P", volume "υ", and temperature "T" is expressed as follows:
$$L = (2/3)\, m \cdot r^2 \cdot \omega = \text{Constant}$$
$$P = \rho \cdot r \cdot \omega^2 = m \cdot \omega^2 / 6\pi r$$
$$\upsilon = (4/3)\, \pi r^3$$
$$T = (2/9k_B)\, m \cdot r^2 \cdot \omega^2$$
Since angular momentum "L" is constant, $P \cdot \upsilon^{5/3}$ and $T \cdot \upsilon^{2/3}$ are expressed as follows:
$$P \cdot \upsilon^{5/3} = (1/2)\,(4\pi/3)^{2/3} L^2/m = \text{Constant}$$
$$T \cdot \upsilon^{2/3} = (1/2\, k_B)\,(4\pi/3)^{2/3} L^2/m = \text{Constant}$$

When υ is altered to V, the equation above is expressed as follows:

$$P \cdot V^{5/3} = \text{Constant}$$
$$T \cdot V^{2/3} = \text{Constant}$$

[End of Proof]

Thus, one could calculate an angular momentum of an ideal gas molecule from pressure and volume of a gas. As shown here, adiabatic process is mathematically calculated using molecular mechanics of the ideal gas under a condition at constant angular momentum.

Exercise

Question One

Find values of molecular radius and rotational frequency of the ideal gas molecule consisting of a single atom with molecular weight "M" at 273.15K under 1atm.

Answer One

Under 1atm (= 101,300 kg·m^{-1}·s^{-2}) at 273.15K, volume "V" of 1mol of an ideal gas is 0.0224m^3. When the radius of a gas molecule is "r", the volume of 1 mol of an ideal gas is expressed as follows:

$$V = (4/3) \pi r^3 N_A$$

Accordingly, <u>molecular radius of an ideal gas is 2.07 nm</u> at 275.15K under 1atm.

From Theorem Ten, pressure is expressed as follows:

$$P = (M/N_A) \cdot \omega^2 / 6\pi r \quad (M/N_A = m)$$

Where M, N_A, and ω indicate molecular weight (kg), Avogadro Number, and angular speed (rad·s^{-1}), respectively. Thus, angular speed of a gas molecule is expressed as follows:

$$\omega = 7.23 \times 10^{12} / M^{1/2}$$
$$\therefore v = \omega / 2\pi = 1.15 \times 10^{12} / M^{1/2}$$

where v indicates frequency (s^{-1}) of rotation of an ideal gas molecule.

Rare gas atoms act as molecules of the ideal gas.

Frequency of rotation of rare gas atoms are calculated as follows:

$$He\ (M = 4.00 \times 10^{-3}) \quad v = 18.2\ THz$$
$$Ne\ (M = 20.2 \times 10^{-3}) \quad v = 8.09\ THz$$
$$Ar\ (M = 39.9 \times 10^{-3}) \quad v = 5.76\ THz$$
$$Kr\ (M = 83.8 \times 10^{-3}) \quad v = 3.97\ THz$$
$$Xe\ (M = 131 \times 10^{-3}) \quad v = 3.18\ THz$$

where THz indicates $10^{12}s^{-1}$.

[End of Answer]

Question Two

Explain the effect of an electric fan wind on temperature of the ideal gas in a room.

Answer Two

Electric fan accelerates translation of air molecules by pressure of the rotating fan. Accordingly, almost all the energy produced by an electric fan is translational kinetic energy of the wind. The pressure of the rotating fan increases temperature of the air to some extent under adiabatic process, while almost all the energy applied to the air molecules by the fan does not increase temperature of the air.

[End of Answer]

Question Three

Answer the most effective way to increase temperature of the ideal gas with a constant energy applied externally to the gas.

Answer Three

Since temperature of a gas is the rotational energy of air molecules, we need to increase rotational energy of air molecules by application of torque on each of them. For this purpose, we need to induce friction between air molecules or between air molecules and room walls at frequency of rotation of air molecules. One possible method is the vibration of walls along its plane at frequency of rotation of air molecules. For example, radiation of infrared light onto the wall increases temperature of the air effectively.

[End of Answer]

Afterword

Here we study on the molecular mechanics of the ideal gas to conclude that rotation of an elastic molecule creates its shape and volume. In the molecular mechanics of the ideal gas, each molecule is a rotational elastic body instead of a translational rigid body in the kinetic theory of gasses. This modification of physical model of molecule may enable us to explain some types of waves in the molecular mechanics.

Gas mediates travelling of sound, light, and electromagnetic waves. The nature of light wave enlightened the model of a gas molecule as mentioned in the "Forewords", while the mechanics of light or optics is still unsolved. It is also unclear how to mediate the electromagnetic wave in the gas molecules. Furthermore, gravitational wave, which reached everywhere through everything in gaseous, liquid, and solid phase, is not understood yet. We have a lot of mysterious questions left to be solved in science. I hope that someone in a young generation will challenge and discover the secret of nature in the future to make difference for human beings.

Finally, I would express my special thanks to my family, Yoko, Kodai, Arisa, and Hiroto for helping and encouraging me to write this pretty book.

At home in Kasukabe

Nov 13th, 2016

Hiroyuki Aizawa

.

www.ingramcontent.com/pod-product-compliance
Lightning Source LLC
Chambersburg PA
CBHW041150180526
45159CB00002BB/765